SNOWFALL!

SNOWFALL!

by Julian May

Published by Creative Educational Society, Inc., Mankato, Minnesota 56001. Published simultaneously in Canada by J. M. Dent and Sons, Ltd. Library of Congress Catalog Card Number: 70-156057. Standard Book Number: 87191-065-9. Text copyright © 1972 by Julian May Dikty. Illustrations copyright © 1972 by Creative Educational Society, Inc. All rights reserved. No part of this book may be reproduced in any form without written permission from the publisher, except for brief passages included in a review. Printed in the United States of America.

Designed by William Dichtl

Cover: Skiers enjoy the beauty of new-fallen snow
CREDIT: H. Armstrong Roberts

Main Title: A lonely fencerow makes a pattern against a winter landscape
CREDIT: Nebraska Game Commission

CREATIVE EDUCATIONAL SOCIETY, INC.
MANKATO, MINNESOTA

Sometimes the snow falls slowly, in fat flakes that stick to everything. Trees and bushes wear thick, white frosting. Telephone wires become furry ropes. Fireplugs and fenceposts have marshmallow hats. This kind of snow is light, fluffy, and fun to play in.

A snowmobiler moves through freshly fallen snow near Dixville Notch, New Hampshire, not far from the Canadian border.

A snowfall is no fun for workers trying to go home. Autos, buses, and
trains are slowed or even brought to
a standstill by thick, slippery snow.

Sometimes the snow falls quickly, blown into great drifts by strong winds. The air is so full of flying flakes that it is hard to see. The snow stings the face. It covers car windshields so that the wipers can hardly clear it away. It piles up on sidewalks and roads and airport runways. Man and his machines cannot move. This kind of snow is heavy, wet, and dangerous.

Sometimes snow falls as a diamond-dust powder. Sometimes it looks like icy needles, or white bits like midget popcorn that bounce on the street. Often the flakes are beautiful crystals, flat with six sides, that look like glass lace under a magnifying lens. When the flake is warmed it turns to a drop of water . . . for that is all snow is: frozen water. The weatherman calls both rain and snow "precipitation," a word that means moisture that falls from the sky.

These beautiful crystals have perfect shapes. Many that fall are broken or irregular, or partly melted and clumped together.

A snowfall in Florida is a rare event. In 1958, students at Florida State University made a sled out of a palm frond. The snow melted quickly.

People who live in very warm parts of the earth may never see a snowfall. . . . Even though clouds high above the tropics may sometimes make snow, the flakes melt and turn to rain long before they strike the earth. Snowflakes melt when the temperature of the air around them goes above the freezing point of water, 32° Fahrenheit.

Most of the northern parts of the world, where the largest cities are, have some snow every year. In North America, snow is rarest in the southern Gulf states, in southern California and Arizona, and in Mexico. Further south, only the highest mountains have snow. And mountain areas in the United States and Canada are likely to have far more snow than lowland places.

The Yellowstone Park area is covered with snow for at least six months of the year. These bison graze on plants near a hot spring, which keeps one small place warm.

U.S. FOREST SERVICE

Rangers push a pipe down into deep snow. It takes out a sample of snow, which is here being weighed to show how much water it contains. The heavier the snow, the more water.

Snowfalls are measured in inches. You can do it yourself by thrusting a ruler down into freshly fallen snow on a level surface. Out West, where melting winter snows are a major source of water, snow rangers measure the snow's depth and also the amount of water in it. Fluffy snow may have less than half as much water as sticky snow.

Cottontail rabbit does not hibernate. It eats mostly green bark in winter.

Ptarmigan turn white in winter to help them escape enemies, such as owls.

The cold winds of winter freeze the earth. Where there is a winter snow-cover, the soil freezes much less. Snow, together with the air it contains, acts like a giant blanket. It is an insulator—a material that keeps heat in and cold out. Snow keeps the soil from freezing deeply. It protects plants. And deep underground, below the frozen top layer of soil, rest many animals that sleep all winter long. Among them are turtles, snakes, frogs, and such mammals as gophers and woodchucks.

The elk will feed mostly on dried grass. Hollow hairs in his coat insulate his body from cold.

Bighorn ram browses on dried grass, twigs, and any other plant food he can find.

Field mice and some other small mammals make tunnels in the snow. Even when the outside temperature is far below zero, the tunnels are a fairly warm 25° F. Animals that are active in winter burrow down into the snow to keep warm. In very cold places, both Eskimos and polar scientists build houses that use snow as an insulation against the winter storms. Man has learned from the animals how to keep warm in the snow.

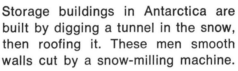

Storage buildings in Antarctica are built by digging a tunnel in the snow, then roofing it. These men smooth walls cut by a snow-milling machine.

An Eskimo snow house, cut from blocks of snow, can be heated with a very small fire because its walls keep the cold air out. Modern Eskimos no longer live in such houses, but a hunter far from home might build one for a shelter.

Many people love snow. The white, winter world is beautiful and clean. Even a dreary, littered place looks good when it is hidden beneath a fresh snowfall. During the daytime, the snow sparkles in the sunlight. At night, crisp black-and-white scenes spread out beneath the moon and stars. And it is exciting when a sudden snow storm changes a well-known place, giving it a mysterious new look.

The city acquires a new look after fresh snow falls.

Skiing is one of the most popular winter sports.

Two New York children speed down a toboggan slide.

Do you like winter sports? When the snow falls,
you can go skiing. Or you can ride a sled or a
toboggan down a slippery slope, or go for a ride on
a snowmobile as it races over a white field or along
a snowy woodland trail. Nowadays, winter
vacations are just as much fun as summer ones.

A warden checks out a hunting camp
deep in the woods. Snowmobiles have
opened up areas that were once cut
off by winter snows. Do you think this
is a good thing?

EVINRUDE MOTORS

But a heavy snowfall is not always fun. For some wild animals, and for many people as well, snow makes life much harder. Deer, with their small hooves, cannot run about on top of the snow as rabbits can. They sink down and have a hard time moving. It is difficult for them to find food. Birds may also find little to eat when the earth is deep in snow.

Deer form winter herds and trample down an area called a "yard." If the herd of deer is large, they may eat all the twigs and other food and begin to starve. These deer are thin and will have to be fed by man.

Stalled cars jam this street in Chicago after a large snow storm. The snow-plow must wait until the cars are pulled out of the way.

People who must work or travel find snow a nuisance or even a danger. Cars, trucks, and buses slip and slide when they try to roll through snow. Airplanes cannot take off or land from snow-packed runways. Trains get stuck in deep drifts. And people trying to do their shopping or get to work stumble along and wish that spring would come.

Snowplows may have to work day
and night to keep the roads open. This
is a pusher-type.

Clearing away the snow is an expensive job. As
soon as the white flakes begin to pile up on the
ground, fleets of snowplows roll out and begin to
remove it from streets, runways, and railroad tracks.
In steep places, sand or cinders are spread to help
wheels get a grip on the ice. Salt and chemicals
are also spread on roadways. They make the snow
melt because saltwater has a much lower freezing
temperature than the plain water of the snowflakes.

Railroads passing over the high mountains have powerful rotary snowplows.

Northern and mountain states that have lots of snow also use rotary plows on the highways.

If the snowfall is really deep, ordinary snow-pusher plows may not be able to clear away the drifts. Another kind of plow, called a rotary, is needed. It chews into snowdrifts with big cutting wheels, then throws the snow high up and to the side, where it is out of the way. But machines cannot help much when the snow falls in narrow or crowded places. Then people must work on the drifts with shovels.

Fourteen inches of snow, falling all
at once, paralyzes a large city. Plows
try to keep the main streets open.
But side streets often must wait.

Usually, the hard-working snow-moving crews keep the highways, tracks, and runways open. But sometimes a single large storm will dump a foot or more of snow in a short time. When a very heavy snowstorm strikes a large city, the plows may not be able to clear it away. All traffic stops. Stores and businesses are closed because the workers cannot leave their homes. The news goes out over television and radio: the big city is snowbound!

Only helicopters and snowmobiles are able to move about in such a storm. They are kept busy rescuing stranded travelers and carrying people to hospitals. When the snow finally stops, the work of digging out begins. It may be many days before the city is back to normal. The great snowstorm may have cost the city millions of dollars in lost time, as well as the expense of clearing away the snow and treating injured people.

After a blizzard, huge drifts may be
piled up fifteen or more feet high.

Some people call every heavy snowstorm a blizzard.
But to a weatherman, a real blizzard is a very
special kind of snowstorm. It has winds of more
than 50 miles an hour, blowing snow, and low
temperatures. Real blizzards are most common in
the northlands of Canada and Alaska, and in
the plains of middle North America. Huge snowdrifts, taller than a house, may be piled up by
blizzard winds. Cattle may die, frozen on the open
range. A real blizzard is almost always a disaster
unless it happens in a lonely place where no one
lives.

All snowfalls, from the lightest powdering of tiny diamond-dust flakes to the worst blizzard, begin in a sky filled with clouds. The highest-flying clouds we see, winter and summer, are made of ice crystals. The lower clouds are formed of very tiny droplets of liquid water. It is these, strangely enough, that produce snow. The clouds often make snow even in summer, but by the time the snowflakes reach the ground, they melt into raindrops.

High cirrus clouds make veil-like wisps over Mount Hood, Oregon, which has snow on it even in summer because it is so high.

Weathermen have known for a long time that only certain kinds of clouds can make snow. But they did not know just how it was made, or why clouds dumped their load of frozen moisture in one place rather than another. The secret was discovered when scientists grew snowflakes in the laboratory. There they could tell exactly how much moisture was in the air, and exactly what the temperature was. They could study the growing snowflakes under a microscope.

Thick, gray "waterdrop" clouds often produce snow in winter. If super-cooled water droplets fall instead of snow, glaze may coat trees and other objects in a glittering jacket of ice.

PUBLICATION ASSOCIATES

A boiling kettle makes a cloud of tiny water droplets. Invisible water molecules (arrow) rush out of the spout. As they cool, they cluster together into a cloud of droplets. Molecules are much too small to be seen even with the most powerful microscope.

Snowflakes grow only when the air contains plenty of moisture. Most of the water floating in the air is in the form of an invisible gas, water vapor. Sometimes the bits of vapor, called molecules, gather together into small droplets of liquid that we can see. These form "water-drop" clouds. But it is important to remember that clouds also have molecules of water vapor that *cannot* be seen—for it is these invisible bits that build into snowflakes.

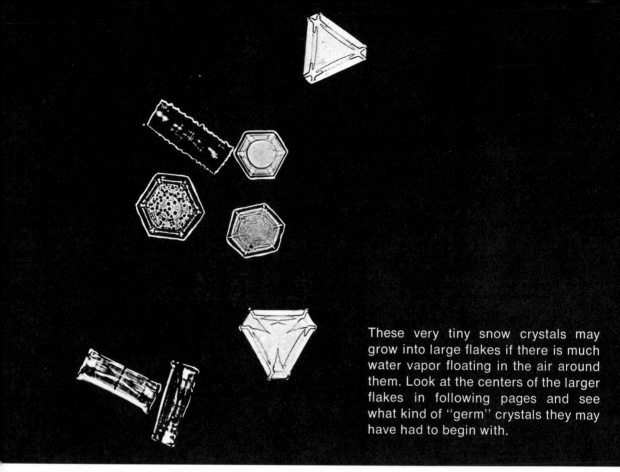

These very tiny snow crystals may grow into large flakes if there is much water vapor floating in the air around them. Look at the centers of the larger flakes in following pages and see what kind of "germ" crystals they may have had to begin with.

Every snowflake is built around a little center of solid matter. This may be a bit of soil, a particle of smoke, or even a speck of dried sea-salt. Molecules of water vapor strike the speck and stick to it. They line up—millions and millions of them—and build a six-sided piece of solid ice called a crystal. The ice crystal has a six-sided shape most of the time because of the "shape" of the molecules of water themselves.

The newborn snow crystals are so small and light that they float without falling to the ground. Unless they grow, there will be no snowfall. A growing snowflake "feeds" on invisible water vapor, gathering countless water molecules as it drifts up and down in the cloud. As it grows, it usually keeps its six-sidedness. But it may change its shape, depending upon the amount of water vapor it is able to gather, and the different temperatures of air layers it passes through.

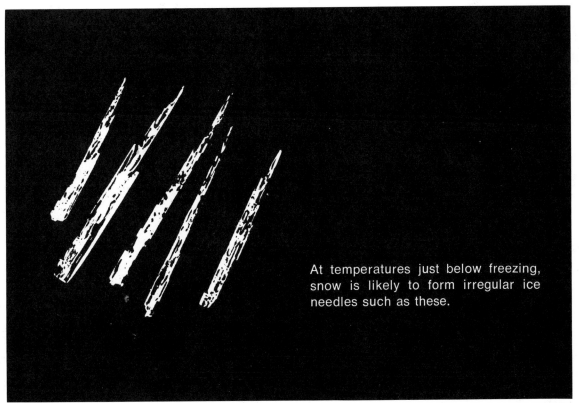

At temperatures just below freezing, snow is likely to form irregular ice needles such as these.

W. A. BENTLEY PHOTOGRAPH

DIFFERENT KINDS OF SNOW CRYSTALS

HEXAGONAL PLATES

CRYSTAL WITH
SECTOR-LIKE BRANCHES

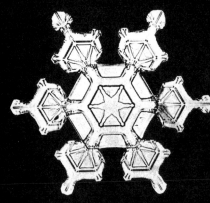

CRYSTAL WITH
BROAD BRANCHES

PLATE WITH
SIMPLE EXTENSIONS

PLATE WITH
SECTOR-LIKE EXTENSIONS

PLATE WITH DENDRITIC
(TREE-LIKE) EXTENSIONS

ORDINARY
DENDRITIC CRYSTAL

DENDRITIC CRYSTAL
WITH PLATES ON ENDS

FERNLIKE CRYSTAL

DENDRITIC CRYSTAL WITH
SECTORLIKE EXTENSIONS

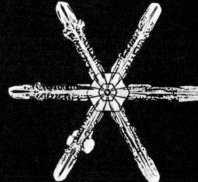

ORDINARY
STELLAR CRYSTAL

STELLAR CRYSTAL
WITH PLATES ON ENDS

DENDRITIC CRYSTAL
WITH 12 BRANCHES

BROAD-BRANCH CRYSTAL
WITH 12 BRANCHES

THREE-SIDED PLATE

MALFORMED CRYSTAL

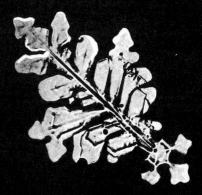

BROKEN BRANCH

Scientists have examined snowflakes and discovered that many of them can be fitted into "families" according to their general shape. Look at snowflakes with a magnifying glass and see how many different kinds you can find. All of these flakes form at "medium" temperature of around 15 degrees F.

COLUMNS

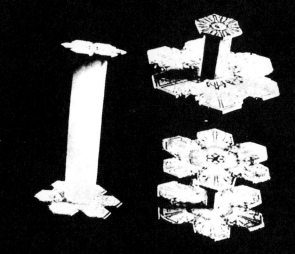

COLUMNS WITH PLATES

RADIATING ASSEMBLAGE
OF PLATES

BULLETS

Very cold air makes bullet-shaped
crystals, or crystals shaped like tiny
columns.

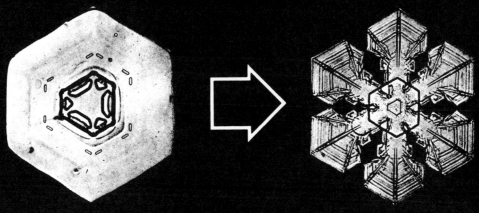

A tiny crystal with a shape like this   . . . may grow into a crystal like this.

The larger and more elaborate a snow crystal is, the slower it must have grown. It is often said that no two snowflakes are alike. This may or may not be. But the beauty of snowflakes is one of the wonders of nature, all the more amazing because they are so small and numerous.

As the snowflake grows large enough to fall, it drops into warmer air. Air high above the earth is usually colder than air near the earth. Entering the warmer air, the snowflake grows faster and faster. This happens because warm air can hold more water vapor than cold air. If you could "squeeze" a bagful of zero-degree air and a bagful of 32-degree air, you would get three times as much moisture from the warm air as from the cold.

RIME-COATED
SNOWFLAKES

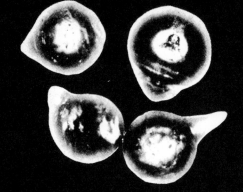

SLEET

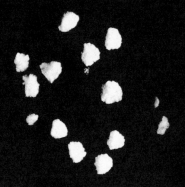

GRAUPEL

Rime-coated snowflakes still show their original form. Graupel or snow pellets look like tiny popcorn. Sleet is glasslike, frozen rain.

Very large snowflakes bump into the water droplets that make up the cloud. These droplets are "super-cooled"—cooled far below the normal freezing temperature of 32°F. When the droplets hit the snowflake, they freeze at once. The beautiful pattern of the snowflake will be hidden by blobs of ice called rime. If many supercooled droplets freeze on the snowflake, it may grow into a fat, white snow pellet. Weathermen call this kind of snow graupel.

A moisture-laden cold wind caused rime to form on these trees.

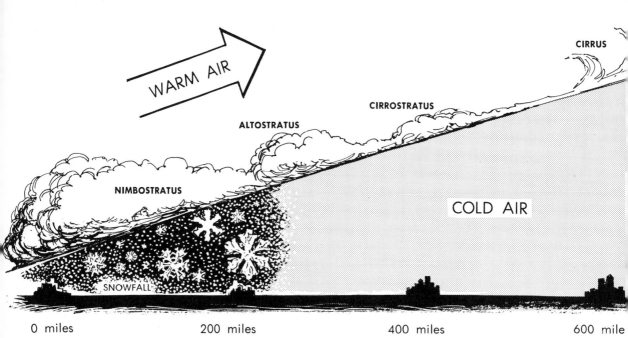

WARM AIR

CIRRUS

CIRROSTRATUS

ALTOSTRATUS

NIMBOSTRATUS

COLD AIR

SNOWFALL

0 miles   200 miles   400 miles   600 mile

PUBLICATION ASSOCIATES

When the weatherman predicts a warm front, he is telling us to watch for the arrival of a mass of warm air, pushing cold air aside. Notice the different kinds of clouds that signal the thickening of the warm-air mass. Low-flying nimbostratus is most likely to make snow.

You have probably noticed that the heaviest snowfall is likely to happen when the air temperature is just around freezing. There is rarely a heavy snow when it is very cold, because cold air can hold such a small amount of moisture. Snow most commonly forms when the air temperature within the cloud is between 23° and –4°F. Of course, the temperature near the earth will usually be warmer. Would you like to be able to tell when it will snow? Watch out for the arrival of warm air during the winter months. Warm winter air is often very moist—and likeliest to produce snow.

When the snow falls to the ground, it immediately begins to change. The beautiful, lacy shapes may be broken as the crystals are tumbled by the wind. If the weather is warm, the snow may melt. But even when it is below freezing, the snow crystals change. Molecules from one part of the flake float off as water vapor without every becoming a liquid. The molecules may "land" elsewhere on the snowflake— or they may float away into the air. Watch a bank of old snow during the winter. You may see it grow slowly smaller without melting. And you will surely see the beautiful fluffy flakes transformed into rounded bits of ice.

True "powder snow," best for skiing, is produced as fallen flakes change into fine grains through slow melting. Several days of sunny, cold weather are needed to transform stiff, newly fallen snow into the best powder.

At warmer temperatures, just around freezing, deep
snow changes to long or cup-shaped bits of ice. A
mass of snow containing these bits is not nearly
as strong as fresh snow. Early springtime in the
snow-covered mountains can be a time of special
danger because of the changes within deep snow-
banks. The ice crystals may suddenly break, causing
a whole mass of snow to begin sliding downward.
This is an avalanche. It can bury a skier, a house, or
even a village.

An avalanche rushes down a mountain in California. It can travel up to 200 m.p.h.

This automobile was carried off the road and smashed by an avalanche in Utah.

The snow patrol keeps skiers off the slopes until dangerous masses of snow can be triggered into small avalanches by means of dynamite or explosive shells.

Avalanches can happen even in cold weather if fresh snow falls upon an old snow field with a crust of slippery ice crystals. The fluffy new snow slides easily over the old. The most terrible kind of avalanche is a great cloud of powder snow moving downhill at speeds up to 200 miles an hour. This kind of avalanche has flattened whole forests. But it is the blast of wind, rather than the snow itself, that does the damage.

Snow rangers are able to tell when avalanches are most likely to happen. They can warn people to get out of the way. But sometimes avalanches happen without warning, and sometimes people are careless about where they ski and walk. Avalanches take many lives each year. But hundreds more are saved by rescue teams of men and dogs, who are trained to seek out and rescue avalanche victims.

People can live for many days while buried in snow because there is usually air mixed in with it. Here the rescue patrol practices probing deep snow with long poles, looking for a dummy "victim."

The dummy is found and "rescued."

U.S. FOREST SERVICE

U.S. FOREST SERVICE

It is harder to protect property from snow slides. Some places have strong walls that turn the flowing snow aside. Towns in the Swiss Alps have special metal "snow fences" that hold back the snow and prevent it from sliding. Railroads in the mountains sometimes build shed-like tunnels where the track runs along a steep slope. Avalanches flow right over the shelter and the trains keep running along.

This kind of track shelter, common in the Canadian Rockies, is expensive and only built where snow and rock slides are very common.

SOUTH DAKOTA DEPT. HIGHWAYS PHOTO BY BOB GAGE

Snow is something that many grumble about. But almost everyone agrees that it makes the drab winter world nicer to look at. Noises are muffled by snow. Ugly things are given a strange, white beauty. The frozen earth is sheltered . . . and later given fresh life when snow melts away beneath the warm rays of the spring sunshine.

ALGIMANTAS KEZYS, S.J.